FOURTH GRADE MATH WORKBOOK: DECIMALS MADE EASY

SPEEDY
PUBLISHING

DECIMAL ADDITION

1. $0.9 + 0.9 =$ _____

2. $1.1 + 1.7 =$ _____

3. $2.3 + 1.3 =$ _____

4. $2.1 + 1.1 =$ _____

5. $0.3 + 1.4 =$ _____

6. $2.2 + 0.3 =$ _____

7. $1.6 + 0.4 =$ _____

8. $0.2 + 0.7 =$ _____

9. $2.7 + 1.5 =$ _____

10. $1.3 + 1.1 =$ _____

11. 0.3 + 0.0 = _____

12. 2.2 + 0.8 = _____

13. 2.6 + 0.4 = _____

14. 1.1 + 0.7 = _____

15. 2.1 + 1.8 = _____

16. 1.0 + 1.7 = _____

17. 1.1 + 0.2 = _____

18. 1.4 + 1.7 = _____

19. 1.5 + 0.6 = _____

20. 2.4 + 1.9 = _____

21. 2.5 + 0.4 = _____

22. 1.6 + 0.5 = _____

23. 1.6 + 1.4 = _____

24. 1.9 + 0.2 = _____

25. 2.1 + 0.8 = _____

26. 0.1 + 1.7 = _____

27. 0.2 + 0.4 = _____

28. 0.4 + 0.9 = _____

29. 1.8 + 0.7 = _____

30. 0.0 + 1.5 = _____

31. 2.2 + 0.2 = _____

32. 1.7 + 1.6 = _____

33. 1.7 + 1.9 = _____

34. 1.3 + 1.0 = _____

35. 0.2 + 0.2 = _____

36. 2.0 + 1.7 = _____

37. 2.9 + 1.3 = _____

38. 2.5 + 1.9 = _____

Decimal addition - 1 digit - LEVEL 2

1. $3.8 + 4.6 =$ _____

2. $4.5 + 2.0 =$ _____

3. $6.6 + 0.2 =$ _____

4. $6.0 + 7.5 =$ _____

5. $7.5 + 8.2 =$ _____

6. $7.4 + 1.4 =$ _____

7. $7.9 + 5.1 =$ _____

8. $9.7 + 6.7 =$ _____

9. $1.0 + 5.7 =$ _____

10. $3.2 + 10.2 =$ _____

11. $1.9 + 6.4 =$ _____

12. $1.1 + 1.9 =$ _____

13. 7.0 + 7.4 = _____

14. 7.2 + 3.7 = _____

15. 8.8 + 1.8 = _____

16. 9.3 + 2.7 = _____

17. 6.7 + 3.5 = _____

18. 4.5 + 4.6 = _____

19. 8.6 + 9.0 = _____

20. 7.4 + 6.2 = _____

21. 9.7 + 1.6 = _____

22. 1.9 + 4.5 = _____

23. 0.8 + 0.2 = _____

24. 7.8 + 2.8 = _____

25. 9.1 + 8.8 = _____

26. 7.9 + 0.5 = _____

27. $8.5 + 0.4 =$ _____

28. $7.6 + 2.7 =$ _____

29. $2.7 + 9.6 =$ _____

30. $8.3 + 6.6 =$ _____

31. $3.2 + 9.6 =$ _____

32. $1.3 + 7.2 =$ _____

33. $0.2 + 1.7 =$ _____

34. $5.2 + 0.9 =$ _____

35. $4.3 + 4.1 =$ _____

36. $4.1 + 8.4 =$ _____

37. $6.7 + 0.3 =$ _____

38. $4.9 + 4.4 =$ _____

39. $3.3 + 2.4 =$ _____

40. $7.2 + 0.3 =$ _____

Decimal addition - 1 digit + 2 digit - LEVEL 3

1. $0.1 + 0.9 =$ _____

2. $0.3 + 0.5 =$ _____

3. $0.9 + 1.7 =$ _____

4. $0.03 + 0.1 =$ _____

5. $1.3 + 1.6 =$ _____

6. $0.9 + 1.3 =$ _____

7. $0.2 + 1.69 =$ _____

8. $1.6 + 1.6 =$ _____

9. $1.5 + 0.7 =$ _____

10. $1.29 + 1.9 =$ _____

11. $1.1 + 1.34 =$ _____

12. $0.4 + 0.07 =$ _____

13. 1.3 + 1.9 = _____

14. 1.0 + 0.0 = _____

15. 0.70 + 1.3 = _____

16. 1.14 + 0.5 = _____

17. 1.6 + 0.8 = _____

18. 1.5 + 0.5 = _____

19. 0.9 + 1.6 = _____

20. 0.0 + 1.1 = _____

21. 0.31 + 0.6 = _____

22. 0.3 + 1.11 = _____

23. 1.9 + 0.4 = _____

24. 1.5 + 1.0 = _____

25. 0.3 + 0.1 = _____

26. 1.2 + 0.58 = _____

27. 0.77 + 1.3 = _____

28. 1.05 + 1.7 = _____

29. 0.1 + 1.37 = _____

30. 1.5 + 0.1 = _____

31. 1.4 + 1.85 = _____

32. 0.1 + 0.5 = _____

33. 1.4 + 1.4 = _____

34. 1.3 + 1.53 = _____

35. 1.2 + 0.31 = _____

36. 1.4 + 0.8 = _____

37. 0.9 + 0.2 = _____

38. 1.7 + 1.2 = _____

39. 1.3 + 0.6 = _____

40. 0.82 + 1.2 = _____

Decimal addition - 1 or 2 digits - LEVEL 4

1. $1.0 + 0.0 =$ _____

2. $1.54 + 0.80 =$ _____

3. $0.03 + 1.8 =$ _____

4. $0.74 + 0.7 =$ _____

5. $1.79 + 0.43 =$ _____

6. $0.3 + 1.49 =$ _____

7. $0.4 + 0.80 =$ _____

8. $0.7 + 1.14 =$ _____

9. $1.5 + 1.54 =$ _____

10. $1.5 + 1.8 =$ _____

11. $0.91 + 0.81 =$ _____

12. $0.3 + 0.9 =$ _____

13. 0.10 + 0.80 = _____

14. 0.7 + 0.9 = _____

15. 0.16 + 1.2 = _____

16. 0.80 + 0.84 = _____

17. 1.42 + 0.0 = _____

18. 0.07 + 1.90 = _____

19. 1.46 + 1.5 = _____

20. 0.85 + 1.16 = _____

21. 1.31 + 1.03 = _____

22. 0.4 + 0.8 = _____

23. 1.7 + 0.31 = _____

24. 0.25 + 1.1 = _____

25. 1.5 + 1.4 = _____

26. 1.9 + 0.89 = _____

27. 1.7 + 0.83 = _____   **34.** 1.56 + 0.83 = _____

28. 1.85 + 0.49 = _____   **35.** 1.19 + 1.36 = _____

29. 1.7 + 0.2 = _____   **36.** 1.7 + 1.08 = _____

30. 1.0 + 0.2 = _____   **37.** 1.6 + 1.98 = _____

31. 0.50 + 0.5 = _____   **38.** 0.95 + 1.7 = _____

32. 1.2 + 0.9 = _____   **39.** 0.2 + 1.9 = _____

33. 1.15 + 0.60 = _____   **40.** 0.70 + 0.25 = _____

DECIMAL SUBTRACTION

1. $3.4 - 2.4 =$ _____

2. $5.5 - 2.3 =$ _____

3. $8.6 - 0.8 =$ _____

4. $0.5 - 0.1 =$ _____

5. $4.6 - 4.3 =$ _____

6. $2.1 - 0.0 =$ _____

7. $4.5 - 0.7 =$ _____

8. $5.7 - 2.2 =$ _____

9. $0.6 - 0.3 =$ _____

10. $5.3 - 4.6 =$ _____

11. 4.6 – 1.0 = _____

12. 10.8 – 2.0 = _____

13. 5.0 – 1.5 = _____

14. 9.6 – 2.4 = _____

15. 8.5 – 4.4 = _____

16. 3.7 – 1.4 = _____

17. 1.2 – 0.1 = _____

18. 5.2 – 4.3 = _____

19. 2.5 – 2.5 = _____

20. 6.9 – 5.8 = _____

21. 3.7 – 0.8 = _____

22. 10.9 – 2.9 = _____

23. 10.3 – 4.0 = _____

24. 9.2 – 2.8 = _____

25. 4.9 – 2.7 = _____

26. 5.8 – 3.8 = _____

27. 3.1 – 2.2 = _____

28. 7.4 – 5.1 = _____

29. 10.6 – 3.4 = _____

30. 5.8 – 4.2 = _____

31. 1.9 – 0.5 = _____

32. 6.0 – 0.0 = _____

33. 8.7 – 2.9 = _____

34. 10.3 – 3.2 = _____

35. 2.8 – 1.5 = _____

36. 9.4 – 2.1 = _____

37. 5.0 – 2.2 = _____

38. 3.7 – 1.0 = _____

1. $0.4 - 0.4 =$ _____

2. $9.5 - 1.8 =$ _____

3. $4.7 - 0.0 =$ _____

4. $1.8 - 1.24 =$ _____

5. $9.8 - 1.5 =$ _____

6. $4.1 - 1.9 =$ _____

7. $2.1 - 1.7 =$ _____

8. $5.86 - 1.5 =$ _____

9. $5.65 - 1.9 =$ _____

10. $7.0 - 1.0 =$ _____

11. $2.47 - 0.8 =$ _____

12. $4.02 - 0.6 =$ _____

13. 9.03 − 1.0 = _____

14. 2.9 − 0.1 = _____

15. 8.6 − 1.7 = _____

16. 1.5 − 0.12 = _____

17. 3.27 − 0.3 = _____

18. 1.9 − 1.6 = _____

19. 4.6 − 1.4 = _____

20. 10.4 − 1.3 = _____

21. 8.91 − 0.0 = _____

22. 4.33 − 1.8 = _____

23. 6.17 − 1.1 = _____

24. 3.0 − 1.1 = _____

25. 3.5 − 0.6 = _____

26. 9.8 − 1.0 = _____

27. 10.48 – 1.7 = _____   **34.** 4.68 – 1.5 = _____

28. 4.71 – 0.4 = _____   **35.** 7.65 – 0.7 = _____

29. 2.10 – 1.9 = _____   **36.** 5.17 – 1.6 = _____

30. 10.3 – 0.7 = _____   **37.** 1.83 – 0.6 = _____

31. 1.11 – 0.8 = _____   **38.** 8.8 – 0.0 = _____

32. 7.66 – 0.0 = _____   **39.** 2.06 – 1.1 = _____

33. 10.0 – 1.1 = _____   **40.** 3.2 – 0.5 = _____

THE MISSING ADDEND

1. _____ + 1.7 = 10.5   **6.** _____ + 7.4 = 8.1

2. _____ + 1.1 = 11.7   **7.** _____ + 0.4 = 5.6

3. 10.7 + _____ = 12.2   **8.** 1.2 + _____ = 7.3

4. 0.5 + _____ = 6.1   **9.** 5.2 + _____ = 7

5. 4.7 + _____ = 5.2   **10.** _____ + 1.8 = 4.5

11. $0.8 + \underline{\hspace{2cm}} = 11.1$

12. $\underline{\hspace{2cm}} + 3.4 = 4.6$

13. $8.2 + \underline{\hspace{2cm}} = 8.7$

14. $\underline{\hspace{2cm}} + 1.6 = 4.5$

15. $\underline{\hspace{2cm}} + 1.4 = 10$

16. $\underline{\hspace{2cm}} + 0.1 = 9.4$

17. $\underline{\hspace{2cm}} + 1.5 = 4.8$

18. $\underline{\hspace{2cm}} + 0.4 = 11$

19. $2.7 + \underline{\hspace{2cm}} = 3.9$

20. $\underline{\hspace{2cm}} + 1.5 = 8.9$

21. $\underline{\hspace{2cm}} + 8.3 = 10.2$

22. $1.9 + \underline{\hspace{2cm}} = 4.3$

23. $3.5 + \underline{\hspace{2cm}} = 4$

24. $\underline{\hspace{2cm}} + 7.0 = 8.7$

25. _____ + 9.6 = 9.9

26. 10.4 + _____ = 12.2

27. _____ + 1.9 = 11.2

28. 8.8 + _____ = 9.5

29. _____ + 0.6 = 8.6

30. 0.1 + _____ = 1.9

31. 0.7 + _____ = 3.9

32. _____ + 8.8 = 8.8

33. _____ + 0.6 = 0.8

34. 5.3 + _____ = 6.4

35. 7.5 + _____ = 8.2

36. 3.8 + _____ = 3.8

37. _____ + 7.2 = 8.6

38. 1.5 + _____ = 10.5

THE MISSING MINUEND/ SUBTRAHEND

Decimal addition - missing minuend/subtrahend

1. $7 - \underline{\hspace{2cm}} = 5.7$

2. $\underline{\hspace{2cm}} - 0.4 = 1.6$

3. $6 - \underline{\hspace{2cm}} = 5.3$

4. $\underline{\hspace{2cm}} - 0.2 = 0.8$

5. $\underline{\hspace{2cm}} - 0.3 = 0.7$

6. $\underline{\hspace{2cm}} - 0.4 = 4.6$

7. $2 - \underline{\hspace{2cm}} = 0.1$

8. $3.9 - \underline{\hspace{2cm}} = 3.4$

9. $\underline{\hspace{2cm}} - 0.5 = 3.5$

10. $\underline{\hspace{2cm}} - 1.5 = 4.5$

11. _____ − 0.0 = 0

12. _____ − 1.1 = 7.7

13. _____ − 0.3 = 8

14. 1.8 − _____ = 0.9

15. 3.5 − _____ = 3.4

16. _____ − 1.3 = 5.7

17. _____ − 0.9 = 9.2

18. 3 − _____ = 1.1

19. _____ − 0.3 = 1.1

20. _____ − 0.7 = 8.3

21. _____ − 0.4 = 5.6

22. _____ − 1.3 = 1.7

23. _____ − 1.9 = 5.1

24. _____ − 0.7 = 3

25. _____ − 1 = 0.2

26. _____ − 0.9 = 0.7

27. _____ − 0.9 = 1.2

28. 4.7 − _____ = 2.8

29. 1.7 − _____ = 1.7

30. 7 − _____ = 6.7

31. 6 − _____ = 4.2

32. _____ − 1 = 0.7

33. 1.4 − _____ = 0.4

34. _____ − 0.6 = 0.6

35. 10.9 − _____ = 10

36. 1 − _____ = 0

37. 3 − _____ = 2.3

38. 6.2 − _____ = 5.5

ADDEND

LEVEL 1

1. 1.8
2. 2.8
3. 3.6
4. 3.2
5. 1.7
6. 2.5
7. 2
8. 0.9
9. 4.2
10. 2.4
11. 0.3
12. 3
13. 3
14. 1.8
15. 3.9
16. 2.7
17. 1.3
18. 3.1
19. 2.1
20. 4.3
21. 2.9
22. 2.1
23. 3
24. 2.1
25. 2.9
26. 1.8
27. 0.6
28. 1.3
29. 2.5
30. 1.5
31. 4.4
32. 3.3
33. 3.6
34. 2.3
35. 0.4
36. 3.7
37. 4.2
38. 4.4

LEVEL 2

1. 8.4
2. 6.5
3. 6.8
4. 13.5
5. 15.7
6. 8.8
7. 13
8. 16.4
9. 6.7
10. 13.4
11. 8.3
12. 3
13. 14.4
14. 10.9
15. 10.6
16. 12
17. 10.2
18. 9.1
19. 17.6
20. 13.6
21. 11.3
22. 6.4
23. 1
24. 10.6
25. 17.9
26. 8.4
27. 8.9

28.	10.3	**4.**	0.13	**21.**	0.91	**38.**	2.9
29.	12.3	**5.**	2.9	**22.**	1.41	**39.**	1.9
30.	14.9	**6.**	2.2	**23.**	2.3	**40.**	2.02
31.	12.8	**7.**	1.89	**24.**	2.5		

32.	8.5	**8.**	3.2	**25.**	0.4	**1.**	1
33.	1.9	**9.**	2.2	**26.**	1.78	**2.**	2.34
34.	6.1	**10.**	3.19	**27.**	2.07	**3.**	1.83
35.	8.4	**11.**	2.44	**28.**	2.75	**4.**	1.44
36.	12.5	**12.**	0.47	**29.**	1.47	**5.**	2.22
37.	7	**13.**	3.2	**30.**	1.6	**6.**	1.79
38.	9.3	**14.**	1	**31.**	3.25	**7.**	1.2
39.	5.7	**15.**	2	**32.**	0.6	**8.**	1.84
40.	7.5	**16.**	1.64	**33.**	2.8	**9.**	3.04

LEVEL 3

		17.	2.4	**34.**	2.83	**10.**	3.3
1.	1	**18.**	2	**35.**	1.51	**11.**	1.72
2.	0.8	**19.**	2.5	**36.**	2.2	**12.**	1.2
3.	2.6	**20.**	1.1	**37.**	1.1	**13.**	0.9

14. 1.6
15. 1.36
16. 1.64
17. 1.42
18. 1.97
19. 2.96
20. 2.01
21. 2.34
22. 1.2
23. 2.01
24. 1.35
25. 2.9
26. 2.79
27. 2.53
28. 2.34
29. 1.9
30. 1.2

31. 1
32. 2.1
33. 1.75
34. 2.39
35. 2.55
36. 2.78
37. 3.58
38. 2.65
39. 2.1
40. 0.95

SUBTRACTION

LEVEL 1

1. 1
2. 3.2
3. 7.8
4. 0.4
5. 0.3

6. 2.1
7. 3.8
8. 3.5
9. 0.3
10. 0.7
11. 3.6
12. 8.8
13. 3.5
14. 7.2
15. 4.1
16. 2.3
17. 1.1
18. 0.9
19. 0
20. 1.1
21. 2.9
22. 8

23. 6.3
24. 6.4
25. 2.2
26. 2
27. 0.9
28. 2.3
29. 7.2
30. 1.6
31. 1.4
32. 6
33. 5.8
34. 7.1
35. 1.3
36. 7.3
37. 2.8
38. 2.7

LEVEL 2

1.	0	**17.**	2.97	**34.**	3.18	**9.**	1.8
2.	7.7	**18.**	0.3	**35.**	6.95	**10.**	2.7
3.	4.7	**19.**	3.2	**36.**	3.57	**11.**	10.3
4.	0.56	**20.**	9.1	**37.**	1.23	**12.**	1.2
5.	8.3	**21.**	8.91	**38.**	8.8	**13.**	0.5
6.	2.2	**22.**	2.53	**39.**	0.96	**14.**	2.9
7.	0.4	**23.**	5.07	**40.**	2.7	**15.**	8.6
8.	4.36	**24.**	1.9			**16.**	9.3
9.	3.75	**25.**	2.9	MISSING		**17.**	3.3
10.	6	**26.**	8.8	ADDEND		**18.**	10.6
11.	1.67	**27.**	8.78	**1.**	8.8	**19.**	1.2
12.	3.42	**28.**	4.31	**2.**	10.6	**20.**	7.4
13.	8.03	**29.**	0.2	**3.**	1.5	**21.**	1.9
14.	2.8	**30.**	9.6	**4.**	5.6	**22.**	2.4
15.	6.9	**31.**	0.31	**5.**	0.5	**23.**	0.5
16.	1.38	**32.**	7.66	**6.**	0.7	**24.**	1.7
		33.	8.9	**7.**	5.2	**25.**	0.3
				8.	6.1		

26.	1.8		15.	0.1	32.	1.7
27.	9.3	**MISSING**	16.	7	33.	1
28.	0.7	**MINUEND/**	17.	10.1	34.	1.2
29.	8.0	**SUBTRAHEND**	18.	1.9	35.	0.9
30.	1.8	1. 1.3	19.	1.4	36.	1.0
31.	3.2	2. 2	20.	9	37.	0.7
32.	0.0	3. 0.7	21.	6	38.	0.7
33.	0.2	4. 1.0	22.	3	39.	8
34.	1.1	5. 1	23.	7	40.	1.6
35.	0.7	6. 5	24.	3.7		
36.	0.0	7. 1.9	25.	1.2		
37.	1.4	8. 0.5	26.	1.6		
38.	9.0	9. 4	27.	2.1		
		10. 6	28.	1.9		
		11. 0	29.	0		
		12. 8.8	30.	0.3		
		13. 8.3	31.	1.8		
		14. 0.9				

ANSWERS

Made in the USA
San Bernardino, CA
10 February 2017